THE LOOK WITHIN

THE LOOK WITHIN

SILO

SAMUEL WEISER

NEW YORK

First published in Spanish under the title
La Mirada Interna in 1973

First English edition published under the title
The Inner Look

Published in 1980 by
Samuel Weiser, Inc.
740 Broadway
New York, N.Y. 10003

Copyright © 1979 by the Synthesis Foundation

ISBN 0-87728-494-6

Printed in the U.S.A. by
Noble Offset Printers, Inc.
New York, N.Y. 10003

FOREWORD

The Look Within is one of those truly exceptional books whose universal style and content prevent one from putting it into any specific national or temporal category; indeed, it is difficult to tell in what time or place it was written. This fact alone would have no real importance were the teachings in this book useful only in certain countries or in a certain historical time. You will notice that this book differs completely from great literary masterpieces which always have a style and content clearly particular to their epoch of origin. *The Look Within*, on the other hand, is a universal book conceived from the deepest and most fundamental of human experiences, and directed to the most essential within the human being. Its value and special characteristics lie in this fact.

In this Foreword we do not intend to comment on the text of the book. It did not seem either necessary or proper to make additional observations about this beautiful book other than the notes added at the end which summarize the subjects and chapters covered. The impatient reader may refer to them should he desire a general scheme of the subjects covered before he enters the actual book.

Although its purpose and style is universal, it is still of some importance to mention the particular circumstances surrounding its birth. These considerations about its origin may perhaps appear disproportionate and far removed from the central theme of the book, but we feel it is important to include them and will put the book in context.

The Look Within was written by Silo in Mendoza, Argentina as the autumn of 1972 was drawing to a close. Just a few years before this, on the 4th of May, 1969, Silo had given his only public explanation of his teachings in a moving oration at the foot of Mount Aconcagua in Mendoza. From the moment of this first speech on, the government prohibited any efforts by Silo to publicly explain his teachings.

Siloism, a teaching which was at this time in the process of being born, suffered from initial persecutions, as have all great doctrines in the past. Those who were imprisoned or martyred in the course of helping develop it overcame many obstacles on the ascending path opened by the initial ideas. Less than five years after *The Look Within* had been written, it had already been translated into all the important languages of the world, bringing to vast numbers of people the teachings that were ironically forbidden in their place of origin. As you can see, this universal book was not written in a great cultural center out of which a great civilization radiates, nor was it written in the pleasant and comfortable surroundings in which most works are conceived.

Perhaps this brief account will, aside from stirring your curiosity, explain the impetus for the unique style that does not reflect any particular place or time and which rather charitably overlooks the oppressive circumstances that surrounded its writing. Who would suspect when reading through these peaceful and profound pages that they were born in such an obscure and barbaric environment?

The Editors

CONTENTS

I.	The Revelation	1
II.	Disposition to Comprehend	5
III.	The Non-Meaning	9
IV.	Dependence	13
V.	Intimation of the Meaning	17
VI.	Sleep and Awakening	21
VII.	Presence of the Force	25
VIII.	Control of the Force	29
IX.	Manifestations of the Energy	33
X.	Evidence of the Meaning	37
XI.	The Luminous Center	41
XII.	The Discoveries	45
XIII.	The Principles	49
XIV.	The Guide to the Inner Road	55
XV.	The Experience of Peace and the Great Passage of the Force	61
XVI.	Projection of the Force	67
XVII.	Loss and Repression of the Force	71
XVIII.	Action and Reaction of the Force	77
XIX.	The Inner States	81
XX.	The Internal Reality	89
	Notes	93

I. THE REVELATION

1. Here it tells how the non-meaning of life is converted into meaning and fulfillment.
2. Here is joy, love for the body, for nature, for humanity, and for the spirit.
3. Here sacrifices, feelings of guilt and threats from the beyond are abandoned.
4. Here the worldly is not placed in opposition to the eternal.
5. Here it speaks of the inner revelation at which all arrive who carefully meditate in humble search.

II. DISPOSITION TO COMPREHEND

1. I know how you feel because I can experience your state, but you do not know how the things I say are experienced. Thus, if I speak without self-interest of what makes the human being happy and free, it is worth your while to try to comprehend.
2. Do not think that you will comprehend by arguing. You may believe that your understanding will become clearer by arguing, but opposition is not appropriate in this case.
3. If you ask me what attitude is appropriate, I will tell you it is to meditate profoundly and without haste on what is being explained.
4. If you reply that you are occupied with more urgent things, I will answer that if your wish is to sleep or to die, I will do nothing to oppose it.
5. Do not argue that you dislike this form of presenting things, because you do not say that of the peel when you like the fruit.

   I state things in the way I consider appropriate, not

in the ways desired by those who aspire to things remote from inner truth.

III. THE NON-MEANING

After many days I discovered this great paradox: those who bore failure in their hearts were able to enlighten their last victory; those who felt triumphant were left behind on the road as vegetals of a dark and diffuse life.

After many days, coming from the darkest of darknesses I arrived at the light, guided not by teachings, but by meditation.

So I told myself the first day:

1. There is no meaning in life if everything ends with death.

2. Every justification for actions, whether those actions are worthless or excellent, is always a new dream which leaves emptiness ahead.

3. God is something uncertain.

4. Faith is something as variable as reason and dreams.

5. "What one must do" may be extensively discussed, but nothing ultimately supports the explanations.

6. "The responsibility" of those who commit themselves to something is no greater than the responsibility of those who do not commit themselves.

7. I move according to my interests, and this makes me neither a coward, nor a hero.

8. "My interests" do not justify or discredit anything.

9. "My reasons" are no better or worse than the reasons of others.

10. Cruelty horrifies me, but neither because of my reaction nor in its own nature is it better or worse than kindness.

11. What is said today by me or by others has no value tomorrow.

12. To die is no better than to live or to never have been born, nor is it worse.

13. Not through teachings, but through experience and meditation I discovered that there is no meaning in life if everything ends with death.

IV. DEPENDENCE

The second day:

1. Everything I do, feel and think does not depend on me.

2. I am variable and depend on the action of the environment. When I want to change the environment or my "I," it is the environment which changes me. I seek the city or nature, social redemption or a new struggle to justify my existence. In each case, the environment leads me to choose one attitude or another. In this way my interests and the environment determine my position.

3. Thus it does not matter who or what decides. I have to live since I am in the situation of living. I say all this, but nothing justifies it. I can make a decision, hesitate, or remain. In any case, one thing is only provisionally better than another, but ultimately there is no "better" or "worse."

4. If someone tells me that one who does not eat dies, I will answer that this is in fact true and that one is obliged to eat, driven by need. This does not imply

that the struggle to eat justifies one's existence. Nor is the struggle bad. The need for sustenance is an individual and collective fact, but it has no meaning in the moment that the last battle is lost. I feel solidarity with the struggle of the poor and the exploited and the persecuted. I feel fulfilled by such identification, but I comprehend that it justifies nothing.

V. INTIMATION OF THE MEANING

The third day:

1. At times I have anticipated events which later took place.
2. At times I have grasped a distant thought.
3. At times I have described places I have never visited.
4. At times I have told exactly what occurred in my absence.
5. At times an immense joy has overwhelmed me.
6. At times a total comprehension has enveloped me.
7. At times a perfect communion with everything has filled me with ecstasy.
8. At times I have broken my reveries and I have seen reality in a new way.
9. At times I have recognized something I was seeing for the first time as though I had seen it before.

All this has made me think. It is very clear to me that

without these experiences I could not have left the non-meaning.

VI. SLEEP AND AWAKENING

The fourth day:

1. I cannot take as real what I see in my dreams, nor what I see in semi-sleep, nor what I see when I am awake but in reverie.

2. I can take as real what I see when I am awake and without reveries. This does not speak of what my senses perceive, because naive and doubtful information is sent by the external and internal senses, as well as by the memory. Rather, it speaks of the activities of my mind when referred to the thought "data." What is valid is that my mind "knows" when it is awake, but "believes" when it is asleep. I rarely perceive reality in a new way, so I comprehend that what is normally seen resembles sleep or semi-sleep.

There is a real way of being awake; it has led me to meditate profoundly on what has been said up to here, and it has opened the door for me so I could discover the meaning of all that exists.

VII. PRESENCE OF THE FORCE

The fifth day:

1. When I was really awake I ascended from comprehension to comprehension.

2. When I was really awake and lacked the strength to continue in the ascent, I was able to extract the Force from within myself. It was throughout my entire body. This energy was present even in the smallest cells of my body, circulating faster and more intensely than the blood.

3. I discovered that the energy concentrated in points of my body when they were in motion, but it was absent when they were motionless.

4. During illnesses, there was either a lack or an accumulation of energy in exactly the affected areas, and many illnesses began to recede if the normal flow of the energy was re-established.

Some peoples knew this, and through diverse procedures which are strange to us today were able to re-establish the flow of energy. They were able to find ways of transferring the energy from one to another which allowed them to produce "illuminations" of comprehension and even physical "miracles."

VIII. CONTROL OF THE FORCE

The sixth day:

1. There is a way of directing and concentrating the Force that circulates through the body.

2. There are points in the body which control what we know as movement, emotion and ideas. When the energy is mobilized in these points, motor, emotional and intellectual manifestations are produced.

3. The states of deep sleep, semi-sleep or being awake arise depending on whether the energy acts more internally or superficially in the body. Surely the halos that surround the bodies or heads of saints or enlightened ones in religious paintings allude to this energetic phenomenon that is occasionally manifested more externally.

4. There is a point of control of the truly awake state and there is a way to bring the Force to this point.

5. When the energy is brought to this point, all the other points of control move in an altered way.

When I understood this and hurled the Force to that

superior point, my whole body felt the impact of a tremendous energy which shocked my consciousness, and I ascended from comprehension to comprehension. I also observed that I could descend to the depths of the mind if I lost control of the energy. Then I remembered the legends about "heavens" and "hells," and I saw the dividing line between these mental states.

IX. MANIFESTATION OF THE ENERGY

The seventh day:

1. This energy in movement could become independent of the body and still maintain its unity.

2. This unified energy was really a second body. Then I remembered the legends of ghosts, of the soul and of the spirit.

3. When the energy left or was separated from the body, its material base, it either dissolved from lack of internal unity or manifested itself externally, as occurred in a number of cases of "action at a distance."

4. I could verify that the exteriorization of that energy was produced from the lower levels of consciousness. In these cases, an attempt against the most primary unity or life of the human being provoked the response of exteriorizing the energy to protect the threatened unity. Thus, in the trances of some mediums whose level of consciousness was low and whose internal unity was endangered, these responses of exteriorization were involuntary and

were not recognized as being self-produced, but were instead attributed to other entities.

The "ghosts" and "spirits" that some people or fortune-tellers felt possessed by were nothing but their own doubles. Their mental state was darkened in a trance through having lost consciousness and control of the Force, and they felt controlled by strange beings who at times produced remarkable phenomena. Undoubtedly many "bedeviled" people suffered such effects.

What was decisive, then, was the control of the Force. This awareness completely changed my conception of daily life, as well as of life after death.

By means of these thoughts and experiences I started to lose faith in death, and I no longer believe in it, just as I do not believe in the non-meaning of life.

X. EVIDENCE OF THE MEANING

The eighth day:

1. The real importance of an awakened life became evident to me.
2. The real importance of eliminating internal contradictions convinced me.
3. The real importance of mastering the Force for the purpose of achieving unity and continuity filled me with joyful meaning.

XI. THE LUMINOUS CENTER

The ninth day:

1. In the Force was the "light" that came from a "center."

2. In the dissolution of the energy was a withdrawal from the center, and in its unification and evolution was a corresponding work of the luminous center.

I was not surprised to find devotion for the sun-god in ancient peoples, and I saw that if some worshipped this star because it gave life to their earth and to nature, others discovered in this majestic body the symbol of a greater reality. Others went even further and received innumerable gifts from this center. These gifts at times "descended" as tongues of fire over the inspired ones, at times as luminous spheres which arrived from heaven, and at other times as burning bushes which appeared before the fearful believer.

XII. THE DISCOVERIES

The tenth day:

Few but important were my discoveries which I summarize in this way:

1. The Force circulates through my body and it is really my life, and the life of all living bodies.
2. There are points of control of its diverse activities in my body.
3. There are important differences between the true state of being awake and other mental states.
4. One can lead the Force to the point of real awakening.
5. The Force is exteriorized as the double, producing extraordinary phenomena. Its exteriorization or "going out of" the body produces those perceptions and manifestations erroneously called "extrasensory." In those cases the data arrive to the senses through energetic variations of the double, or arise from the centers towards the world, also due to variations of the double.

6. The double can increase its unity by breaking contradictions, and then the human being progressively achieves the state of being awake.

7. The cohesion of the double begins through the internal representation of the "light," and therefore at the level of the system of images where these representations are possible. Achieving this level of the system of images depends, in turn, on the integration of the mental contents.

8. The double can consolidate itself through unifying activity in daily life, or upon activating the luminous center with the Force. These conclusions made me realize the germ of a great truth in the prayer of ancient peoples. It was lost in rituals and external practices which did not allow them to reach a level where they could develop the look within, which realized with perfection, puts one in contact with the luminous source.

Finally, I noticed that my "discoveries" were not discoveries, but were due to the inner revelation at which all arrive who, without contradictions, search for the light in their own hearts.

XIII. THE PRINCIPLES

Different is the attitude towards life and things when inner revelation strikes like lightning.

By following the steps slowly, meditating on what has been said and on what has yet to be said, you may convert the non-meaning into meaning.

It is no longer a matter of indifference what you do with your life. Your life, subject to laws, is exposed to possibilities which you may choose.

I do not speak to you of liberty, I speak to you of liberation, of movement, of process. I do not speak to you of liberty as something static, but rather of liberating yourself step by step, in the same way as one who approaches a city becomes liberated from the distance already traveled on the road.

Then "what one must do" does not depend upon distant, incomprehensible and conventional morals, but upon laws: laws of life, of light, of evolution.

Here are the said "Principles" you must observe if you want to achieve internal unity, lost since the beginning of the centuries:

1. To go against the evolution of things is to go against yourself.
2. When you force something towards an end you produce the contrary.
3. Do not oppose a great force; retreat until it weakens, then advance with resolution.
4. Things are well when they move together, not in isolation.
5. If day and night, summer and winter are fine with you, you have surpassed the contradictions.
6. If you pursue pleasure, you enchain yourself to suffering. But as long as you do not harm your health, enjoy without inhibition when the opportunity presents itself.
7. If you pursue an end you enchain yourself. If everything you do is realized as though it were an end in itself, you liberate yourself.
8. You will make your conflicts disappear when you understand them in their ultimate root, not when you want to resolve them.
9. When you harm others you remain enchained, but if you do not harm anyone you can freely do whatever you want.
10. When you treat others as you would have them treat you, you liberate yourself.
11. It does not matter in which faction events have placed you. What matters is for you to comprehend that you have not chosen any faction.

12. *Contradictory and unifying acts accumulate within you. If you repeat your acts of internal unity then nothing can detain you.*

You will be like a force of Nature when it finds no resistance in its path. Learn to distinguish what is a difficulty, a problem, an inconvenience, from what is a contradiction. If the problems move or incite you, the contradiction immobilizes you in a closed circle.

Whenever you find great strength, joy and goodness in your heart, or when you feel free and without contradictions, immediately be internally thankful. When the contrary happens, ask with faith, and the gratitude which you have accumulated will return transformed and amplified in benefit.

XIV. THE GUIDE TO THE INNER ROAD

Only if you have comprehended what has been explained before are you in a condition to liberate the Force. If you have not fully comprehended the previous points, which moreover should be converted into your daily way of facing life, it is convenient that you meditate on them until they become clear and practical. If they continue to be obscure after long and patient meditations, search for those who precede you on the road so that they may help you in your evolution, as you would help others with less experience should they come to consult you.

Now follow with attention what I am about to explain to you, since it deals with the internal landscape that you may encounter when working with the Force, and with the direction you can imprint on your mental movements.

"On the inner road you may walk darkened or luminous. Attend to the two ways that unfold before you.

"If you let your being hurl itself towards dark regions, your body wins the battle and it dominates. Then sensations and appearances of spirits, of forces, of remembrances will arise. This way you descend more and more. Here dwell Hatred, Vengeance, Strangeness, Possession, Jealousy, and the Desire to Remain. If you descend even more, you will be invaded by Frustration, Resentment and all the dreams and desires that have brought ruin and death upon humanity.

"If you impel your being in a luminous direction, you will find resistance and fatigue at each step. There are things to blame for this fatigue of the ascent. Your life weighs, your memories weigh, your previous actions impede the ascent. The climb is made difficult by the action of your body which tends to dominate.

"In the steps of the ascent you will find strange regions of pure colors and unknown sounds.

"Do not flee purification which acts like fire and horrifies with its phantoms.

"Reject startling fears and disheartenment.

"Reject the desire to flee towards low and dark regions.

"Reject the attachment to memories.

"Remain in internal liberty with indifference towards the dream of the landscape, with resolution in the ascent.

"The pure light dawns in the summits of the great mountain chain and the waters of the thousand colors flow among unrecognizable melodies towards crystalline plateaus and pastures.

"Do not fear the pressure-of-the-light that moves you further from its center, each time with increasing strength. Absorb it as though it were a liquid or a wind. Certainly, in it is life.

"When you find the hidden city in the great mountain chain you must know the entrance. But you will know this in the moment in which your life is transformed. Its enormous walls are written in figures, are written in colors, are *sensed*. In this city are kept the done and the to-be-done. But to your inner eye, the transparent is opaque.

"Yes, the walls are impenetrable for you!

"Take the Force of the hidden city.

"Return to the world of dense life with your forehead and your hands luminous."

XV. THE EXPERIENCE OF PEACE AND THE GREAT PASSAGE OF THE FORCE

1. Completely relax your body and quiet your mind. Then imagine that a transparent, luminous sphere descends towards you and ends up lodging in your heart. You will recognize the moment when the sphere ceases to appear as an image and transforms itself into a sensation inside your chest.

2. Observe how the sensation of the sphere slowly expands from your heart towards the outside of your body while your breathing becomes more ample and deep. When the sensation reaches the limits of your body, you can stop the whole operation there and register the experience of internal peace. You can remain in it for the time you consider adequate. Then, contract the previous expansion arriving as in the beginning at your heart in order to detach yourself from your "sphere," and conclude the exercise calm and satisfied. This work is called "the experience of peace."

3. On the other hand, if you would like to experience the great passage of the Force, instead of contracting the expansion you have to augment it, letting your

emotions and your whole being follow it. Do not try to pay attention to your breathing; let it act by itself, while you continue the expansion beyond your body.

4. I must repeat this: your attention, in such moments, must be *on the sensation of the sphere that expands*. If you cannot achieve this, it is advisable that you stop until you have mastered it with time. In any case, if you do not produce the passage, you will experience the peace.

5. If you have carried everything out correctly with inner purity and without sudden fears, you will begin to experience the great passage. Your hands and body will begin to be "electrified." Later, all of your body will receive progressive undulations, and in a short time images and emotions will erupt forcefully. Then let the passage be produced without fear, since there will always be someone close to you, who will gently calm you if the need arises.

6. Upon receiving the Force, you will perceive the light and new colors and sounds, but what is important will be the experience of amplification of the consciousness.

7. When the great passage becomes intolerable or excessive for you, you must end this state by imagining or feeling that the sphere contracts and leaves you as it arrived in the beginning.

8. It is better for several people to work together in the great passage. Otherwise, when working alone, one may fall into a state of trance. The consciousness in trance is called "crepuscular," which reveals it to

perfection. In the range of the crepuscular one recognizes: hypnosis, spiritualism, the effects of drugs, and in general all phenomena that overcome the control of the clear and ascending consciousness, while producing the great passage. Almost all "extrasensory" experiences are produced from the crepuscular. Distrust such manifestations, and as told in the legends, consider them the "temptations" that the saints suffered in their evolution.

9. Even if you have worked and carefully observed these recommendations, you may not have been able to produce the great passage. This indicates a lack of internal unity. But do not convert this into a focus or preoccupation, since continued work tends to surpass this with time. Meanwhile, each new experience of peace will be of increasing fullness.

XVI. PROJECTION OF THE FORCE

1. Those who have achieved the passage and who direct it for their evolution have obligations to those on the road behind them. It is useful for those who come after the older ones to gather wisdom and experience from them.

2. In many cases those who have control of the Force can project it to others who, in spite of their efforts, do not achieve the great passage. This is done in meetings where everyone has this strong and sole desire. When the great passage occurs, it is sufficient that contact be maintained among those in different conditions for the Force to be displaced. This projection has been known since ancient times as the "laying on of hands."

3. The Force can be projected to others and also to objects particularly adequate to receive and conserve it. I hope it will not be difficult for you to understand the function fulfilled by the sacraments in many religions, and the significance of sacred places, and of the priests and pontiffs supposedly "charged" with the Force. Later all these were con-

verted into fetishes without profound significance. When some objects were worshipped with faith in the temples and were surrounded with ceremony and ritual, surely the energy accumulated through repeated prayers "returned" to the believers. Underlying what later became superstition—still conserved by ignorant peoples—was the real intuition of the existence of the Force and the ability to transfer it.

4. Let us speak now of the case where you have already experienced the action of the Force. The sudden circulation of the Force in the organism mobilizes the points of control of which I spoke earlier. Then, little by little, one begins to perceive a profound and positive change in the general attitude of those who carry out this work.

5. Each new great passage is a reinforcement of the change being experienced. This will be healthy if it is produced each time with greater expansion of the consciousness. An opposite result clearly shows that operations must be suspended.

6. I suppose your case is the best, and therefore I can now recommend that after the experience of peace or of the Force, you conserve the beneficial sensation for some time.

XVII. LOSS AND REPRESSION OF THE FORCE

1. The greatest discharges of the energy are produced by uncontrolled acts which are: excessive sexuality, imagination without restriction, immoderate small-talk, uncontrolled curiosity, and exaggerated perception—to look, listen, taste, etc., in a frantic and useless way. But you must recognize that many do these things in order to discharge tensions which would otherwise be painful for them. Considering this, and seeing the function of such discharges, you will agree with me that it is not reasonable to repress them, but rather to give order to them.

2. As for sexuality, you must correctly interpret this: this function must not be repressed because you will create mortifying effects and internal contradictions. Sexuality begins and ends in its act, and in no way must it continue to affect the imagination or urge the search for a new object of possession. It becomes evident that if external stimuli or fantasies exaggerate the function of sex, it is excessive.

3. It is clear that the recommendations of chastity and abstinence were at one time related to the Force;

however the instructors who put their attention on this did not explain it in a repressive manner, but in an over-elevated way.

4. Later the control of sex by a certain social or religious "moral" served purposes that had nothing to do with evolution, but rather the contrary.

5. The Force overflowed towards the crepuscular in repressed societies, and cases multiplied of the "bedeviled," "witches," sacrilegious people, and criminals of all kinds who rejoiced in suffering and the destruction of life and beauty. In some tribes and civilizations, the criminals were found among both those who executed and those who were executed. In other cases, all that was science and progress was persecuted because it was opposed to the irrational, to the crepuscular and to the repressed.

6. The repression of sex still exists in certain primitive peoples, as well as in others considered "advanced civilizations." Although the origin of this situation is different in the two cases, it is evident that in both instances the destructive manifestation is great.

7. If you ask me for more explanations, I will tell you that in reality sex is sacred in itself and is the center from which all life and creativity springs, just as all destruction arises when the function of sex is not resolved.

8. Never believe the lies of the poisoners of life when they refer to sex as something despicable. On the contrary, there is beauty in it and not in vain is it related to the best feelings of love.

9. Be careful then, and consider sex a great marvel that one must treat with delicacy without converting it into a source of contradiction or a distintegrator of the vital energy.

XVIII. ACTION AND REACTION OF THE FORCE

It was explained before in one of the Principles: "Whenever you find a great strength, joy and goodness in your heart, or when you feel free and without contradictions, immediately be internally thankful."

1. "To be thankful" means to concentrate positive moods that are associated with the internal sensation or image of the sphere that you know. The positive mood is linked to the sensation or representation and by evoking the same representation of the sphere in unfavorable situations, the positive mood associated with it will arise. Since this mental "charge" is elevated by the work with the Force, it is capable of displacing the negative emotions that certain situations impose.

2. Because of all this, what you ask for will return amplified in benefit from your interior, provided that you have made contact with the Force and have accumulated numerous positive states within yourself.

XIX. THE INTERNAL STATES

You must now acquire sufficient perception of the different internal states in which you may find yourself throughout your life, and particularly throughout your evolutionary work.

1. The first state, in which the non-meaning is prevalent (the one we mentioned at the beginning), is known as the state of simple and diffuse Vitality. Everything is directed by physical needs, but these are often confused with contradictory desires and images. In this state there is obscurity in your motives and in your activities. You remain vegetating in this state, lost among undefined forms. From this point, you can evolve through only two possible ways: the way of Death or the way of Mutation.

2. The way of Death puts you in the presence of a dark and chaotic landscape. Ancient peoples knew this passage and almost always located it "underground" in the abysmal depths. The instructors also visited this kingdom, to later "resurrect" in luminous levels. Understand well that below Death lies diffuse Vitality. Perhaps the human mind re-

lates the mortal disintegration to later phenomena of transformation, and perhaps it also associates the diffuse movement with what happens before birth. If your direction is the ascent, Death signifies a break with your former stage. Through Death one arrives at other states.

3. Arriving at the next state you find the refuge of Regression. Two roads open from here: one is the road of Remorse, and the other is the road of Death which served for the ascent. If you take the first one, it is because your decision tends to break with your past life. If you go back along the road of Death, you destroy your evolutionary possibility and fall again into the depths.

4. As I told you, there is another path of escape from the state of Vitality: the path of Mutation. If you choose this road, it is because you want to emerge from your painful state but are unwilling to abandon some of its apparent benefits. Therefore, it is a false road known as the "road of the left hand," the word "left" meaning "twisted." Many monsters have emerged from the depths through this tortuous passage. They wanted to take the heavens by assault without abandoning the hells, and consequently have projected endless contradiction into the normal world.

5. I will suppose that by ascending through the kingdom of Death and through your conscious Remorse, you have already arrived at the dwelling of the Tendency. You cannot rest very long; you can barely stop. Two narrow ledges sustain your dwelling: Conservation and Frustration. Conservation

is false and unstable. Walking through it you are deluded with the idea of permanence, but in reality you descend rapidly. If you take the road of frustration, your ascent is painful, but the only-one-not-false.

6. After failure upon failure, you can arrive at the next dwelling called "the dwelling of Deviation." Take care with the two roads you now have before you: either you take the road of Resolution that carries you to the lodging of Generation, or you take the road of Resentment which makes you descend again towards Regression. Here you face the dilemma: either you choose the labyrinth of conscious life and do it with Resolution, or you go back to your previous life through Resentment. There are many who, unable to surpass themselves, here cut off their possibilities forever.

7. But you who have ascended with Resolution now find yourself in the first shelter known as "Generation." Here you have three doors: one is called "Fall," another is called "Intent" and the third is called "Degradation." The "Fall" carries you directly to the depths, and only an external accident can move you towards it. It is difficult for you to choose that door. On the other hand, the door of "Degradation" carries you indirectly to the abysses, going back over roads in a sort of turbulent spiral in which you continually reconsider all that has been lost and all that has been sacrificed on the altar of an unknown god. This examination of consciousness that carries you to Degradation is certainly a false examination in which you underesti-

mate and disproportionately evaluate some of the things which you compare. You compare the effort of the ascent with those "benefits" that you have abandoned. But if you look more closely at things, you will see that you have not abandoned anything for this reason, but for other reasons. The Degradation begins when one falsifies the motives and were apparently foreign to the ascent. I ask you now: what betrays the mind? Perhaps the false motives of initial enthusiasm? Perhaps the difficulty of the undertaking? Perhaps the false memory of sacrifices that did not exist, or that were made for other reasons? I say this and I ask you now: your house burned down some time ago. Because of this you decided on the ascent. Now do you think that because you ascended your house burned down? Have you considered what is happening to the surrounding houses? There is no doubt that you must choose the middle door, that of Intent.

8. Go up the stairway of Intent and you will arrive at an unstable dome. From here, move through a narrow, winding hallway known as "Volubility," until you arrive at a space that is wide and empty like a platform and has the name "open-space-of-the-energy."

9. In this space you may be frightened by the deserted, immense landscape and the terrifying silence of that night transfigured by enormous and immobile stars. Here, right over your head, you will see fixed above the firmament the insinuating form of the Black Moon. Here you must await

dawn patiently and with faith because nothing bad can happen to you if you remain calm.

10. It could happen that in such a situation you might want to arrange an immediate exit by your own inventiveness. If that happens you might grope your way anywhere instead of awaiting the day with silence and faith. But you must remember that all movement here is *false* and is generically called "Improvisation." If you forget what I tell you now and begin to improvise movements on your own, be certain that you will be dragged by a whirlwind along paths and among dwellings to the darkest depth of dissolution.

11. How difficult it must be for you to comprehend how the internal states are enchained to one another! If you could see what inflexible logic the consciousness has, you would notice that one who blindly improvises in these matters fatally begins to degrade and to degrade oneself. Then, feelings of Frustration arise. Later one falls into Resentment and finally arrives at Death, in which forgetfulness overcomes everything that one had once managed to perceive.

12. If you manage to reach the day in the open space, the radiant sun will rise before your eyes, illuminating reality for the first time. Then you will see that a Plan lives in everything that exists.

13. It is unlikely that you will fall from here unless you voluntarily decide to descend to the dark kingdoms to carry the light.

I should speak no longer of these truths. Without experience they serve only to deceive by transferring to the realm of the imaginary what may actually be carried out.

May what has been said so far be of use to you, you who come from the faraway non-meaning.

If what has been explained here should not be useful for you, to what could you object, what could you put above this teaching, since in any case, nothing has any basis and reason in your state of existence—an existence close to the image of a mirror, to the sound of an echo, to the shadow of a shadow.

On the contrary, rejoice if a luminous cord has descended into the world of darkness. Rejoice! But remember well that we have brought the real word of redemption which says: "Only you can redeem yourself."

XX. THE INNER REALITY

1. Reflect on these considerations. In them you will have not only intuitions of allegorical phenomena and landscapes of the external world, but also real descriptions (non-allegorical) of the mental world.

2. Do not believe that the "places" you pass through in your journey are those the double touches in its separation from the physical body. Such confusion often darkened profound religious teachings, and even today it is believed that the "heavens," "hells," "angels," "demons," "monsters," "enchanted castles," "jungles," "remote cities," and the rest have visible reality for the enlightened. The same prejudice, but with opposite interpretation, has taken hold of skeptics without wisdom who took these things to be "illusions," or "hallucinations" suffered by feverish minds.

3. It is important to repeat: comprehend that all this deals with real mental states, although they are symbolized by "objects" without existence in themselves.

4. Consider what has been said and learn to discover

the truth behind the allegories that on some occasions deflect the mind, but on others translate realities impossible to grasp without such representations.

When people spoke of the cities of the gods which numerous heroes of different peoples wanted to reach, when people spoke of paradises in which gods and humans lived together in a transfigured original nature, when people spoke of falls and deluges, great inner truth was told. Nevertheless, when all this was told and placed outside of the mind, it was an error or a lie.

The heroes of this age fly towards the stars. They fly through regions previously ignored. They fly outward from their world, and without knowing it they are impelled towards the internal and luminous center.

NOTES

The Look Within consists of twenty chapters, each of which is divided into parts.

The important themes of the book may be grouped in the following way:

A. The first two chapters are introductory, and present the intention of the instructor, the attitude of the one who learns, and the way to implement this relationship.

B. In chapters three through twelve, the more general teaching is developed and explained in ten "days" of meditation.

C. The thirteenth chapter marks a shift from the general teachings to the correct conduct and attitude in facing life.

D. The remaining chapters explain in detail the internal work.

The order of the themes is the following:

Chapter I, The Revelation—This explains the object of the book, that is, the conversion of the non-meaning into meaning through the inner revelation which is achieved by meditating.

Chapter II, Disposition to Comprehend—This explains the correct attitude towards the comprehension of the teachings.

Chapter III, The Non-Meaning—This explains that there is no meaning in life if everything ends with death.

Chapter IV, Dependence—What human beings do depends not on themselves, but rather on the environment.

Chapter V, Intimation of the Meaning—The experience of "suggestive" mental phenomena gives the human being the intimation of a meaning in life.

Chapter VI, Sleep and Awakening—This speaks of differences between levels of consciousness (sleep, semi-sleep, vigil with reverie and full vigil) and related perceptions of reality and of the external senses, the internal senses and the memory.

Chapter VII, Presence of the Force—This explains the increase of comprehension in vigil and the need for a greater internal energy for the ascent to other levels of consciousness. This energy or Force accumulates and moves in the body.

Chapter VIII, Control of the Force—This speaks of the centers and the energy and the depth or superficiality of the energy is related to the levels of consciousness.

Chapter IX, Manifestations of the Energy—The existence of paranormal phenomena is related to the externalization of the energetic double. The control or lack of control of the energy is important. Conscious control of the Force is related to the increase of internal unity and to the level of consciousness. Externalization occurs from the lower levels of consciousness.

Chapter X, Evidence of the Meaning—The need to break internal contradictions in order to achieve unity and continuity is considered.

Chapter XI, The Luminous Center—The energy is linked to the internal allegory of the "luminous center," the phenomena of internal integration and the "ascent" towards the light. The internal disintegration is registered as a withdrawal from the light.

Chapter XII, The Discoveries—The following are discussed: the circulation of the energy or Force, the Centers, the Levels of consciousness, the conducting of energy towards high levels of consciousness, and the externalization of the energy. Internal disintegration and integration are related to contradiction or unity. The nature of the Force is represented as "light." The consolidation of the double depends on unifying activities or procedures of internal work. There are examples of various peoples with respect to these discoveries. These "discoveries" are seen as a manifestation of one's own mental phenomena.

Chapter XIII, The Principles—The Principles convert the non-meaning into meaning in daily life. They do not come from a conventional moral but from laws proper to the functioning of the mind. The observance of the Principles has practical utility in daily life and allows one to continue evolving and gain internal unity.

Chapter XIV, The Guide to the Inner Road—This describes the internal representation of the phenomena that accompany the paths of "descent" and "ascent." It indicates the direction to impart to the mental processes in the work with the Force and the state in which the experience should conclude.

Chapter XV, The Experience of Peace and the Great Passage of the Force—Procedures for achieving the experience of peace and the great passage of the Force are outlined.

Chapter XVI, Projection of the Force—The word "projection" is used here in the sense of external manifestation of the energy. This projection lies within the field of the paranormal phenomena. Leaving this aside since such phenomena are always questionable, it explains the ability of repeated work with the Force to mobilize the activity of the centers and unblock psychic fixations. This gives unity and the integration of internal contents in one's internal experience.

Chapter XVII, Loss and Repression of the Force—This explains energetic discharges and sex as the productive center of energy. The repression of sex orients mental phenomena in the direction of "descent" in which crepuscular states of consciousness are manifested. The exaggeration of sex is a loss of energy.

Chapter XVIII, Action and Reaction of the Force—The association of representations with emotional charges is explained. The evocation of the image, which was recorded with positive emotional states, brings up or "returns" the associated emotional states. "Being thankful" is a technique of associating images with emotional states and is to be used in daily life.

Chapter XIX, The Internal States—This refers to mental situations in which those who are in the internal work may find themselves. These situations are allegorized in such a way that one can grasp their meaning. It is also explained how these situations are linked one to another and do not appear in isolation.

Chapter XX, The Internal Reality—In conclusion, it is explained that the mental processes are linked to allegorical representations of the external world; that they have mental reality although they have no external existence.

Information about practical works based on this book can be obtained from the following locations of the Synthesis Foundation Communities:

I. *USA*

    1. New York: 150-37 Village Road, Parkway Village, Flushing Queens, N.Y. 11432. Tel: [212] 969-1225

    2. Washington, D.C.: 4607 Connecticut Avenue, N.W., #813, Washington, D.C. 20008. Tel: [202] 686-0219

    3. Chicago: 7007 N. Sheridan #318, Chicago, Ill., 60626. Tel: [312] 973-0244

    4. Portland, Ore.: 8650 S.W. Canyon Lane, Portland, Ore. 97225. Tel: [503] 297-3967

    5. Sacramento, Ca.: 1826 H. Street, #C, Sacramento, Ca. 95814. Tel: [916] 443-1257

    6. San Francisco: 1459 17th Avenue, San Francisco, Ca. 94122. Tel: [415] 566-9392

    7. Los Angeles: 9701 Wilshire Blvd. Suite 811, Beverly Hills, Ca. 90212. Tel: [213] 275-0095

    8. Houston: 2208 Portsmouth, Houston, Texas 77098. Tel: [713] 520-6067

II. *CANADA*

    1. Toronto: P.O. Box 1191, Station "B", Downsview, Ontario M3H 5V6. Tel: [416] 633-4595

III. *ENGLAND*

    London: 31 Fairholme Road, London, W14.

IV. *AUSTRALIA*

    Sydney: P.O. Box 1649, North Sydney, N.S.W. 2060. Tel: [02] 921581

V. *MALAYSIA*

    Kuala Lumpur: 2 Jalan 12/10-C, Petalong Jaya, Selangor. Tel: 51851

VI. *HONG KONG*

    D/2 Concord Court -18, 13 South Bay Close, Repulse Bay.